Ugly-Cute

Ugly-Cute

WHAT MISUNDERSTOOD ANIMALS
CAN TEACH US ABOUT LIFE

JENNIFER McCARTNEY

HarperCollins*Publishers*

To all the ugly-cute souls out there living their best (adorable) lives and who inspired this book.

HarperCollins*Publishers*
1 London Bridge Street
London SE1 9GF

www.harpercollins.co.uk

HarperCollins*Publishers*
1st Floor, Watermarque Building, Ringsend Road
Dublin 4, Ireland

First published by HarperCollins*Publishers* in 2022

13 5 7 9 10 8 6 4 2

Copyright © HarperCollins*Publishers* 2022

Jennifer McCartney asserts her moral rights
as the author of this work.

Text by Jennifer McCartney
Cover and interior design by Jacqui Caulton

A catalogue record for this book is
available from the British Library

ISBN: 978-0-00-852705-1

Printed and bound in Latvia

MIX
Paper from
responsible sources
FSC™ C007454

This book is produced from independently certified FSC™ paper
to ensure responsible forest management.

For more information visit: www.harpercollins.co.uk/green

"I would warn you that I do not attribute to nature either beauty or deformity, order or confusion. Only in relation to our imagination can things be called beautiful or ugly, well-ordered or confused."

—**Baruch Spinoza**

Contents

A very optimistic introduction

> "Optimism—the doctrine or belief that everything is beautiful, including what is ugly."
>
> **—Ambrose Bierce**

From the pug to the anteater, some members of the animal kingdom have, historically, been unjustly labeled as not so attractive. Uncute, even. Or, to put it more unkindly: ugly. These misunderstood creatures have been denied the love and appreciation showered on some of their more adorable counterparts (looking at you, koala bears).

And yet, if there's anything we've learned over the past few years, it's that we're all worthy of love, we're all designed exactly how we were meant to be, and some of us are even pretty darn cute, too (once you take a second look). Perhaps there's beauty everywhere—we're just often too judgmental or conditioned by society's boring, bland standards to see it.

Ugly-Cute will restore these lovable creatures to their rightful place in the animal kingdom, and in our hearts. They deserve it all. All the love, and all the rewards that unchecked consumerism has to offer: stuffed animals, stickers, cartoons, and t-shirts. Soon, if the ugly-cute animals have their day in the sun, you'll be able to wear a tee that proudly proclaims you're "Totally Aye-aye," or tell a grandparent that their face is as cute as that of a blobfish.

There is no true universal beauty standard, of course. We're all wonderful, beautiful creatures, whether we're covered in hair or totally hairless, bland and brawny, or colorful and curvy, happiest underwater or baking in the desert sun. We didn't always think this way, though! We've come a long way as a society. These days, thankfully, all the models trying to sell us stuff more often reflect the diversity of our beautiful planet. But we still have a lot to learn, and who better to teach us than the ugly-cute animals who are just waiting for the world to give them a chance? You'll find that once you do, they have a lot to teach us—and they're happy to help. They've been waiting, after all, for us to open our eyes to their wonderfulness.

So, here we are at last. It's never too late to learn. Join me, as we explore the wonderful world of ugly-cute animals, and maybe we'll even learn a little something. And once you're done reading all about these wonderful, beautiful creatures, consider joining the Ugly Animal Preservation Society. It's a place full of like-minded folks (blobfish fanatics) who know that our worth isn't defined by our looks.

From well-known lovable uggos like pugs and sloths, as well as obscure weirdos like the star-nosed mole and the monkfish, enjoy your introduction to these ugly-cute animals and the ways in which we can learn from them!

What is ugly-cute?

Kama muta and
the science behind the
ugly-cute attraction

> "Like most qualities, cuteness is
> delineated by what it isn't."
>
> **— William S. Burroughs**

The famous author William S. Burroughs was talking about cats here—animals he most definitely thought were cute. But he's also correct. You can't have ugly without the cute. So what is cute exactly?

Cute is something attractive, pretty, or endearing. You know, panda bears. Wide-eyed babies. Baby pandas' eyeballs—etc. In a study called "Too Cute for Words," published in *Frontiers in Psychology*, researchers even have a name for how we feel upon seeing something cute: *kama muta*. This is an emotional response described as "being moved or touched, heartwarming, nostalgia." Also known as "the feels," or "I want to squish your face." Some languages have a word for this cuteness reaction—in Finnish, it's *heltyä*, and in Estonian it's *heldinud*. It's thought that *kama muta* happens when we're exposed to tiny helpless things like, well, human babies. This helplessness triggers our caregiver response and we feel protective. No surprises there.

On the flip side, ugly generally means something unpleasant or repulsive in appearance. It's also suggestive of violence, as in, "Oooh, this situation could get ugly."

It seems straightforward: Ugly on one end of the spectrum and cute on the other. But *how do we really know* which is which? Is our response to cuteness hardwired and forever? Or is it socially conditioned? Basically, is it possible that the things we deem ugly one year can become cute the next? Well, yes. Anyone in art and fashion knows this to be true. Remember men's mustaches? The denim your mom wore? Floral wallpaper? These things went from universally cute to ugly and back to cute again as the decades went by. Their essential properties never changed. Only our view of them did. See how easy it is to gain a new perspective? Just remember everything is fluid and nothing is for certain. And sometimes all it takes is a second look, or a bit of time to appreciate what we're looking at. So with that in mind, let's meet some ugly-cute animals.

The rare beauty of ugly-cute

One thing that many of these ugly-cute animals have in common is that they're found in tiny geographical areas. In one single lake. On one small island. In one bit of swamp on the coastline of a faraway place. That's neither ugly nor cute—that's beautiful. Remember, not everything is binary. We contain multitudes.

The ugly-cute philosophy and the BLOB Method

"If you think something is ugly, look harder. Ugliness is just a failure of seeing."

—Matt Haig, *The Humans*

We have a lot to learn from ugly-cute animals—how to survive, how to thrive, and how to just be ourselves. We can also learn how to do all these things without the constant praise and admiration we so desperately crave (and often deserve!). Everyone loves applauding a panda bear for giving birth, but does a dung beetle ever get a handclap for its excellent sense of direction? So here's a quick way to remember how to keep those ugly-cute vibes going in your own life. We present to you—the BLOB Method. Because there's nothing nobler, more chill, and more wonderful in the natural world than the misunderstood—and very cool—blobfish.

Be weird. The planet is full of nonsense. Things that seem inefficient. Beings that make you think, why does this even exist? The fact is, weird things exist. Weird things thrive, even. And no matter how put together we seem on the outside, the fact is we're all a bit weird. We're furry in places we shouldn't be. We look a bit odd. Our behavior confounds the scientists. We've got habits the world can't explain. And yet—it's all good. We're out here, living our lives, making friends, getting work, and finding love. Long live the weirdos. We come to be appreciated, eventually.

Love indiscriminately. Why not? Life is short. We've just enough time to love as hard as we can in as many ways as we can before passing on. Love your new socks. Love the sunlight glinting off your phone screen. Love all the various bits and bobs of your body that we're taught we shouldn't love. Love thy neighbor. You get the idea.

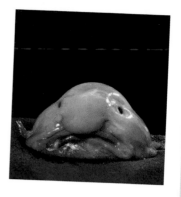

Open up. We're closed to lots of things in life, and lots of times this is due to fear or anxiety. We're closed to new experiences, new people, new situations. And that's understandable. It's easier to close up. Sometimes, to be sure, we're closed to protect ourselves. It's good to be cautious. But like everything in life, this response is best in moderation. So while it's easier said than done, try an open mind next time you experience something new and unexpected. There may be beauty there.

Bury the hatchet. Some things in life are legit ugly. Ugly experiences, ugly memories, ugly interactions that we hold on to because we can't forget. Letting that ugliness in and letting it live with us for too long, though, can be counterproductive. If you can, let the ugliness go. Accept it. Process it. And embrace what happens next. Moving on is a beautiful thing. A sloth in Costa Rica told me that once. Sloths tend to move on much more slowly than most, of course, so take your time.

Quiz:

Find your ugly-cute animal muse

Quiz

Are you an Aye-aye or an Armadillo?

1 Choose one of the following sentiments to describe yourself:

   a) Tiny but mighty.
   b) Expect the unexpected.
   c) Weird and slightly wild.
   d) Loyal and listening.

2 Your friendship style is:

   a) Mostly loner. Well-written books are your friends.
   b) Selective. I choose my friends carefully and allow them to hang around me.
   c) Dramatic. Full of artistry, top restaurants, and late nights.
   d) Friends for life. Long talks with people you've known since you were five. And with your barista. And the mailwoman.

3 When someone wrongs you, your reaction is:

   a) Retreat, and wait it out. This will all blow over soon.
   b) I'm going to remember this for the next nine years, then get my revenge.
   c) A quick public duel will ensure this is all sorted out quickly.
   d) Let me think about this from their perspective for a moment. I don't want to misinterpret anything.

4 Your dream job is:

    a) Park ranger or lighthouse operator.
    b) Quality assurance for a pillow company.
    c) Art gallerist in Paris or top chef in Dubai.
    d) Couples therapist or hostage negotiator.

Answers:

Mostly As: Pink fairy armadillo. Soft and delicate but also tough (they've got a little suit of armor, after all), the pink fairy armadillo person is comfortable in their own skin, and happy living life in the great outdoors, unfettered and free.

Mostly Bs: Hairless cat. Unique, unexpected, but also keeps people on their toes. The hairless cat person runs their world with confidence—on their own schedule and on their own terms.

Mostly Cs: Anteater. These animals served as muses for many Surrealist artists back in the day. The anteater person is adventurous, bold, a little bit strange, and always keeps people guessing.

Mostly Ds: White-faced saki monkey. One of the most loyal animals around, if this monogamous monkey is your muse, that means you exude trustworthiness, you're a good listener, and you're loyal for life. Friends (and strangers) love coming to you for a good chat!

A menagerie of ugly-cute animals

Pink Fairy Armadillo

Pink fairy armadillos *sound* lovely; they were blessed with a name a bit more charitable than the blobfish. Unfortunately, they do look a bit like a baby chick wearing a snakeskin cape. If the chick had claws...

FACT: The smallest species of armadillo in the world and found only in Argentina, these ugly-cute animals have a soft, thin, and flexible pink shell—unlike the tough armor we're used to seeing on an armadillo. Their underbellies are covered in silky, yellowish-white fur and their tails look like a spatula. So, a bit weird, perhaps. They *also* like to live alone and die very quickly in captivity. As quickly as four days after being captured. So let's leave them be, shall we? They don't need our judgment and they're happy by themselves. A nice reminder that not every species is desperate for love and attention (looking at you, peacocks).

Aye-aye

An aye-aye is a lemur with large eyes and extremely long fingers that's found in Madagascar. Its teeth, according to scientists, never stop growing. And that's not even the oddest thing about them.

FACT: It's one of only two animal species that use the percussive foraging method. It knocks on a tree, listens for grubs, gnaws a quick hole (with its extra-long teeth), then reaches in (with its extra-long digits) and scoops out its dinner. Even more astonishing is that each finger has a special use. Its third finger is just used for tapping, while its fourth is used for hooking the grubs. It also has a sixth finger, which is used for orchestra conducting, probably. In some cultures, aye-ayes have a poor reputation and have even been accused of using their long fingers to puncture the aortas of sleeping locals. Sometimes things unique and wonderful *can* seem scary. But they likely won't murder you.

Sloth

As a society, we've collectively agreed that sloths are pretty cute. They have the covered-in-fur and sweet lil' face that make them prized in the animal kingdom (and lucrative for plush animal and yearly calendar producers). But you may be surprised to learn that this wasn't always so.

FACT: Judgmental French naturalist Georges Buffon described sloths as "the lowest form of existence" when writing about them in 1749. "Slowness, habitual pain, and stupidity are the results of this strange and bungled conformation," he wrote. "One more defect would have made their lives impossible." Well, a hearty heave-ho to old Buffon, because while sloths are indeed slow, vulnerable to predators, and in need of lots of rest (about 15–20 hours a day), they're also survivors. The oldest types of sloth were around millions of years ago, and here they are today, going about their business, albeit slowly—the only thing that's changed is how we see them. They knew we'd come around eventually.

"I am overcome by my own amazing
sloth … Can you please forgive
me and believe that it is really
because I want to do something
well that I don't do it at all?"

— Elizabeth Bishop

Star-nosed Mole

There are many ways in which one could describe the star-nosed mole's appearance, and none of them are kind. In place of a face, it has ... a sort of nose with twenty-two fleshy appendages. Lucky for the mole, these appendages make it one of the world's most highly sensitive animals.

FACT: The *Guinness Book of World Records* named the star-nosed mole as the world's fastest-eating mammal. Its starred bits are so sensitive they can identify, catch, and eat food in milliseconds. The number and receptivity of these appendages are of great interest to science. How does it catch prey so quickly? Can it "see" with its nose? Does it sense the electrical field of its prey? Or does it process information by touch only? Science isn't sure. What's clear, though, is that sensitivity comes in many different forms. Which is interesting to contemplate. What if instead of sensing stimuli with our five separate organs, everything was mushed into one super sensor? You could taste music and hear tangerines. Life is infinitely weird and full of possibilities—and galaxies. Like ours. Which includes the star-nosed mole. Lucky us. Knowing what you do now, doesn't their curious pink nose in the shape of a star seem sort of adorable, somehow?

Wombat

This protected Australian species is short-legged and muscular with rodent-like teeth. It also has a pouch on its butt, and it expels distinctive poops in the shape of a cube. Wombats use their scat to mark their territory and attract a mate, so the fact that their scat is stackable and less likely to roll is an inherent evolutionary advantage. A madness, indeed! (And possibly not the most enticing dating profile, unless you're a female wombat.) In 2020, scientists at the Western Australia Museum discovered that wombats (along with some other marsupials and mammals) are biofluorescent. That means they glow in the dark! Why? That is still being figured out ...

FACT: These new-to-England beasts made quite a splash among the Victorians—the English poet Rossetti became a bit obsessed with them after seeing one in the Regent's Park Zoo, in London, in 1847. He bought two wombats and kept them as pets. In fact, he wrote a poem about the one that didn't immediately die in his care.

Oh how the family affections combat
Within this heart, and each hour flings a bomb at
My burning soul! Neither from owl nor from bat
Can peace be gained until I clasp my wombat.

Another fact is that a group of wombats is known as a wisdom. Or a mob. What you call a group of them probably depends on whether you've just stepped in anything cube-shaped on your walkabout. Rossetti *almost certainly* called them a wisdom.

"The Wombat is a Joy, a Triumph,
a Delight, a Madness!"

—**Dante Gabriel Rossetti**

Sucker-footed Bat

Bats don't have a wonderful reputation, thanks to Bram Stoker's *Dracula* and every bit of folklore from every country over the last couple of centuries. But there's no time like the present to become enlightened to the wonders of the bat community.

FACT: These real-life Spider-Man creatures use a secretion in their feet and wrists to attach themselves to stuff. This sticky substance is controlled by the bat, letting them walk upside down—or whichever way they want. Unlike other, less sticky bats, they like to sleep curled up in giant leaves. This is so they don't fall on their heads when their little foot secretions dry up while they snooze. Cute. Unless you brush up against a bedtime leaf full of dozens of them. In which case, you'd be forgiven if you let out a concerned yelp. Turns out even the most enlightened among us can still get a bit startled by the unexpected. Deep breaths.

Sun Bear

Generally, people *love* to love bears—teddy bears, Paddington Bear, Winnie-the-Pooh, etc. They're cute! So where did it all go wrong for the sun bear? Partly it's just a crowded market. Pandas, grizzlies, brown bears—they get a lot of press. Partly it's because they look a tad different from their more familiar bear relatives. They've got a long lolling tongue, a short little snout, and a pigeon-toed walk. They're often spotted walking on their hind legs, which gives them a kind of Sasquatch vibe. If you saw one strolling upright through the forest, you'd be forgiven for posting about your Bigfoot encounter.

FACT: This tree-loving black bear is the smallest member of the bear family—and also the rarest. They're so-named because of the creamy fur patch on their chest that looks like the rising sun. They've got extra-long tongues, which can be a bit shocking or a bit impressive, depending on your proclivities. It's this hopeful little patch of fur (along with their lolling tongues) that cements their place in the official bear hall of cuteness. Like if someone you knew wore a rainbow T-shirt every day—broadcasting optimism and sunshine to all who are lucky enough to encounter them. Which is admirable, because sun bears are also endangered due to hunting and loss of habitat—soon they may be as rare (or as nonexistent, depending on your level of skepticism) as their Sasquatch lookalikes. Shine on, little sun bears. May you live to see another sunrise.

Tapir

Tapirs look like a cross between an anteater and a hog. And a rhinoceros. With a teeny, elephant-type trunk. It's one of those rare animals that can only be described by comparing it to other things. Unfortunately, they have a reputation for being a bit dim—but to be fair there's still a lot that science doesn't know about them. Like why they love to dive, swim, and walk underwater. Imagine, on your next snorkeling trip, seeing a giant pig happily clomping its hooves along the lakebed floor while snuffling up aquatic plants and you'll get a sense of how weird the tapir can appear to the uninitiated.

FACT: Baby tapirs are quite firmly on the cute side of things—they resemble newborn fawns with spotted, stripy fur and a sweet little face. It's thought they're multi-colored to better blend in with the under-brush, keeping them safe from predators while they're teensy. Another way tapirs escape predators is by taking to the water and using their snout as a snorkel. Fully submerged, that little bit of exposed nose can keep them well stocked with oxygen until the threat passes. It gives new meaning to the phrase "keeping your head above water." If tapirs can do it, so can you. Whatever keeps you alive is worth a shot.

Anteater

These fluffy, long-nosed, and long-tongued animals look like walking feather dusters. It's no wonder the Surrealists took to them. Salvador Dalí was once spotted walking one on a leash in Paris. This is a bad idea, of course, owing to the fact that anteaters very much dislike people, other animals, and even other anteaters. With their long, sharp claws they've even dispatched a few unlucky humans who couldn't take the hint. Death by anteater sounds ... surreal, doesn't it?

FACT: Anteaters eat (you guessed it) ants, along with termites and other insects. They do this by lapping them up with their two-foot-long tongues covered in small hooks for gripping—they can lick and flick them up to 2.5 times per second. Marvelous! (There's another reason they belong to the family of animals known as Vermilingua, which means worm tongue!) Lacking teeth, the insects go straight to the anteater's stomach, where they're ground up by its tummy muscles. When the tongue is extended it does look a bit like an expended party favor—you know the ones, where the string pops out of the canister with a bang. Which is kind of the whole anteater's vibe. Surprise! Worm tongue in the house! Let's celebrate how ridiculous it is that I exist. Time to pop the bubbly, etc. A short while later, you'll be embracing the adorable anteater and telling them how they've been your favorite all along.

"I went to the zoo once and saw this thing they call an anteater. That was quite enough for me."

— **Thomas Pynchon**

White-faced Saki Monkey

The pale-headed or white-faced saki monkey is so-named due to the appearance of the male of the species—they sport a white head with a black body and a fetching little mustache. The females are more brownish. They both, however, have little human-like ears that are a bit startling. All the better to hear you with, my dear. They're found in South America and spend most of their time in trees, hanging out with their one true love and chatting up a storm.

FACT: These monogamous primates mate for life. To keep things spicy with their chosen monkey partner, they talk and sing together as a way of bonding. This is why a karaoke bar is such a good idea for a date or friend outing. It's a socially acceptable time to be terrible at singing, and everyone will admire your effort while feeling closer to you romantically and platonically due to science reasons. I don't know, just ask the monkeys. They seem to have it all figured out and apparently they're great listeners.

Yeti Crab

Are these terrifying? Slightly. They're ghostly, underwater fur creatures with claws. But don't they also look a bit like a stuffed animal? They're cute, right? You kind of want to give it a hug. Good luck finding one, though. These hairy-armed weirdos live in hydrothermal vents deep in the ocean.

FACT: They were (imaginatively) dubbed "hairy lobsters" (*Kiwa hirsuta*) when scientists discovered them deep off the coast of Easter Island in 2006. Their hair traps bacteria, which they eat, supposedly. Or maybe they eat shrimp; science isn't sure. All we do know is that deep, deep under the sea, in the frigid cold, at depths inhospitable to 99 percent of living things, a bunch of furry crabs are hanging around some heat vents with their friends, surviving against all odds.

Pug

Pug owners think pugs are magnificent, which is a bit like saying your mom thinks you deserve a raise. The owners, and your mom, are a bit biased. And there's nothing wrong with that. It's nice to be appreciated even when the wider modern world hasn't quite realized your worth.

FACT: Pugs originated in China and have been the dog of choice for various royal houses over the centuries, going all the way back to the Song Dynasty. Queen Victoria was a pug-lover, along with Marie Antoinette, and so were many Tibetan monks who kept them as pets. Goya and William Hogarth painted pugs (the latter's pug portrait hangs in the Tate) and Jane Austen wrote about them, although rather unkindly. The dogs are renowned for their genial disposition—the American Kennel Club describes the breed as "even-tempered and charming." So if you find snuffling and snorting charming, as many pug owners do, then perhaps this is the breed for you. Now can someone get this dog a tissue?

"She was a woman who spent her days in sitting, nicely dressed, on a sofa, doing some long piece of needlework, of little use and no beauty, thinking more of her pug than her children."

—Jane Austen, *Mansfield Park*

Quiz:

Find your ugly-cute love match

Quiz

Are you attracted to Magnificent Frigatebird types? Or are you more of a die-hard Dung Beetle?

1 When it comes to love, you're looking for:

   a) Hard-working types with a good job, no frills necessary.
   b) Glorious pantomime, drama, splendor!
   c) A good time, someone chill.
   d) Loyalty, adoration, regular walks.

2 Your perfect date would be:

   a) Weightlifting, or some other form of exercise.
   b) A day at sea on a fancy sailboat.
   c) Beach day! Or maybe beach week. You've got nowhere to be.
   d) A walk in the park, maybe a game of frisbee, and some cuddles.

3 You'd move anywhere for love, but preferably:

   a) Cairo, Egypt.
   b) Key West, Florida.
   c) Monterey, California.
   d) Shanghai, China.

Answers:

Mostly As: Dung beetle (also known as the sacred scarab in Egypt). The strongest insect in the world, these tried-and-true marvels represent the hardest-working, most dependable sort. Whatever happens, these types can handle it.

Mostly Bs: Elephant seal. These creatures are known for being sociable but chill. They're happy just to hang out with you and your friends. Preferably in the sun.

Mostly Cs: Magnificent frigatebird. Incredible stamina, these birds can stay aloft for more than two months without landing. Good luck with that!

Mostly Ds: Pug. These loyal types are enthusiastic, loving, and unconditionally devoted. Find yourself a match like this and you'll be set for life.

Axolotl Salamander

Wide heads, no eyelids, and fluffy gills that look like pink plants growing from behind their ears. This is the critically endangered axolotl salamander found only in a single lake on the edge of Mexico City (although there are tons of them bred in captivity for reasons that will become clear).

FACT: The axolotl is an aquatic amphibian that lives and breathes underwater. Unless, as scientists somehow discovered, you expose them to iodine, after which they develop more muscle mass, grow eyelids, and start using their dormant lungs. In other words, they transform into air-breathing amphibians that can live on land. As you can imagine, this metamorphosis makes them incredibly interesting things to study. Oh, and they can also regenerate damaged limbs and regrow parts of their brain and eyeballs. So it's possible these remarkable little salamanders—once a staple of the Aztec diet—could hold the key to some really wild medical breakthroughs in the near future. Not bad for something whose name translates to "water monster." An image of the tiny but mighty amphibian now graces Mexico's 50-peso note.

Monkfish

The thing about appreciating the "monstrous" monkfish is that it's all about angles—as any experienced selfie-taker understands. When they're hanging out in their element on the ocean floor, they can look a bit silly when viewed from above. Think Muppet ready to sing with a gruff voice and floppy eyebrows. Or a bit of lichen with eyes. When you spot one on a fishmonger's table, however, it's a manifestation of your worst nightmare. With giant prehistoric mouths, rows of teeth, and flat, slippery fish bodies, these bottom-feeders are also known as monster fish, frog fish, and sea devils. They can grow to massive sizes. They've even been known to eat sea birds and otters. Yikes! Let's look again at that watery Muppet.

FACT: Despite their controversial appearance, monkfish are considered a delicacy in many countries. In Japan, steamed monkfish liver—known as *ankimo*—is served at high-end sushi spots. These sustainably caught fish are also popular in Nordic countries and in some more-enlightened regions of North America. So, once again, how you view the monkfish is all about perspective. It's an adorable ocean dweller that looks perfectly natural in its ocean-floor habitat. It's an expensive delicacy served by the world's best chefs. Or, if you're not as open-minded, it's the dinosaur fish of death. How you choose to see it is up to you. But, might I suggest, life will be a lot more pleasant if you take the former view.

Proboscis Monkey

These bulbous-nosed primates are known for their fleshy face append-ages, which are sometimes so large they interfere with mealtimes. Imagine having to set aside your own nose to enjoy a cup of Earl Grey? That's just a day in the life of these ugly-cute legends (although they prefer mangrove leaves to tea leaves). Found exclusively on the island of Borneo, in Southeast Asia, these unique monkeys also have slightly webbed feet and love to swim. If you catch one sitting down, they look a bit like a benevolent troll from an old-world folktale.

FACT: Proboscis monkeys have been known to indulge in a belly flop here and there—they've been spotted jumping into water from heights of up to 50 feet. It's thought that the thwacking sound they make when they hit the surface might startle any crocodiles in the area, giving the monkey time to get its backstroke on. They can also dive and swim underwater for up to 65 feet at a time. Their aquatic abilities are rare in the primate world—likely a result of their unique island habitat, which sees flooding throughout the year. And they look great swimming, by the way. Natural talents. So the next time you hit up a pool or beach, channel some of that monkey-leaping joy and try your hand at a belly flop. No matter how silly we might feel in our birthday or bathing suits, and no matter what we might look like to the outside world, the fact is, life is short. Make the leap! I promise you'll look cute doing it.

Aquatic Scrotum Frog

This is a salacious name, we admit. The proper name for the scrotum frog is the Titicaca water frog. Not quite as descriptive, though, is it? This amphibian is found only in South America's Lake Titicaca basin in the Andean highlands between Bolivia and Peru. It was also, sadly, a contender for World's Ugliest Animal a few years back—a tongue-in-cheek designation awarded by something called the Ugly Animal Preservation Society.

FACT: These are large frogs with lots of excess skin that gives them a wrinkled appearance. Sort of like, well, you know ... Or, if you're being more generous, they look like they're rocking a very voluminous fashionable coat that one might see on the runway but not understand. Or a toddler that has disappeared for a moment too long and returned wearing every article of her dad's clothing. However, it turns out that this excess skin is very important to its survival. Because Lake Titicaca is the highest lake in the world, the sorts of species that survive in this low-oxygen, high-UV environment have to be pretty tough. It's thought that because the frog breathes through its skin, the increased surface area due to these extra folds allows it to maximize its oxygen intake. Pretty amazing. Anyway, being named after some very natural dangly bits shouldn't really be considered an insult in these enlightened times, surely? We can all be proud of the cute little bits and bobs that make up the human body—as it turns out, everything has a purpose.

Emu

Emus have wings but they can't fly. They can run fast and kick pretty hard—hard enough to kill. Lethal force isn't generally something that co-exists with cuteness; no one is worried about a baby seal kneecapping them, for example. But the emu retains its status as an ugly-cute bird based purely on how weirdly and adorably it's put together. Big old fluffy body, powerful bird legs, and a long ostrich-like neck topped with a curious little head. What's not to love? Plus, they lay bright-green eggs, which is neat. (The eggs, incidentally, are laid in a nest built by the male emus, who then incubate the eggs and raise the chicks.)

FACT: This flightless Australian bird can weigh up to 120 pounds. It's able to store fat quite easily and can lose close to 50 percent of its body weight while looking for more food. Some enterprising soul once figured out that each emu holds around 2 to 3 gallons of oil—which means emu soaps, lotions, and other wellness products are a popular treat Down Under. Luckily for the emu, there are a lot of them and they're not endangered—in some areas they're even considered pests. So, lather up. Or dig in; emu steaks are a healthy alternative to beef—the protein-rich meat is low in cholesterol and high in iron and vitamin C.

"I once traveled to Adelaide on Emu Airways. I was 5,000 feet up in the air when someone pointed out to me that emus can't fly."

—**Billy Connolly**

Blobfish

"What would be ugly in a garden constitutes beauty in a mountain."
—**Victor Hugo**

Blobfish have rather unfairly become the face of the ugly animal world. As you might suspect, their public debut via a series of unfortunate photographs wasn't entirely fair. These deep-sea fish belong to a family called "fathead sculpins" and are found in the waters of Australia, Tasmania, and New Zealand.

FACT: The blobfish's unfortunate moniker is the result of snapshots taken while it was literally a fish out of water. Not having any bones, the fish doesn't photograph so well, but the fishermen who originally photographed it didn't realize this. All they saw was a lump of jelly. And so the "blobfish" was born. The stud in the stunning portrait opposite is the iconic Mr Blobby, taken by research scientist Kerryn Parkinson. Gelatinous, sweetly smiling, and, yes, blobby, the fish never had a chance at being taken seriously. Underwater, however, at depths of up to 4,000 feet, the fish look pretty grand—the water pressure gives them a bit of shape and form. Underwater, they look like themselves. Not so blobby now, eh?

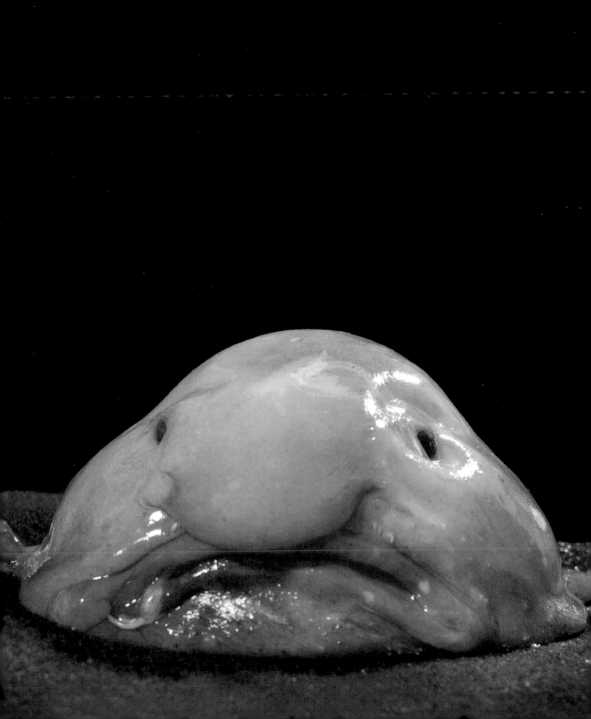

Hairless Cat

The hairless cat is a victim of poor branding. It's known far and wide as *the hairless cat*, which is an accurate description, to be fair. But its proper name is the Sphynx, which is much more dignified. This particular breed is the result of a natural mutation found in a single kitten born in 1966 in Toronto, Canada. For some reason, the owner of this hairless kitten thought, *Adorable! I must have more!* Today, as a result of that breeding process, the hairless cat, with its suede-like skin and wrinkled face, is among the most popular breeds in the world.

FACT: Sphynx require lots more care than their hairy counterparts. They're at risk of sunburn, and they must be given weekly baths to keep the cat oils to a minimum. They can also get cold, so owners in chilly climates tend to fit them out with little coats. Obviously, this is where the cuteness factor comes in. Cats generally hate clothing, so we're not accustomed to seeing them in Christmas sweaters or wearing puffer coats, but that's the joy of a hairless cat. They're outfitted for their own adorable protection! This practice probably helps with the continuation of the species, in all honesty. Right now, someone, somewhere, is falling in love with a Sphynx cat for the very first time, planning out nine lives' worth of glorious wardrobe changes and the accompanying photoshoots.

Elephant Seal

The scientific name of the elephant seal is *Mirounga*, which sounds aptly like the chilled-out greeting one seal would use to another. Both species of elephant seal are huge, with males maxing out at 20 feet and 4.5 tons; however, they do their lounging in very different locales. Northern seals adore the vibes of California and Baja California, while their less touristy cousins the southern seals live in sub-Antarctic and Antarctic waters. During breeding season, male seals use their good looks to collect a harem of up to 50 females. Once a year, the California State Park of San Simeon Beach hosts up to 18,000 elephant seals crowded onto a 4-mile stretch of beach, accompanied by such a racket of bellows and grunts, you'd think you'd stumbled onto the losing tryouts for *America's Got Talent*.

FACT: Nature-lovers drive for miles to see elephant seals hanging out on the California coast during mating season. Just like us, these gigantic ocean mammals have a favorite beach, where they like to hang out with their friends and look for love. Imagine tourists driving for miles to watch you hang out at the beach with your friends? You're there, in your element, lolling about, drinking some wine from a box, then here come hundreds of excited fans—delighted by your total embrace of the sun, your total chillness, your *vibe*. That's the life of an elephant seal. They don't win any awards for their nimbleness or delicate ways, but they don't need your dumb awards, they're having a relaxing beach day with their friends.

Kākāpō

"Sorry, but this is one of the funniest things I've ever seen.
You are being shagged by a rare parrot."

—Stephen Fry

This large, flightless, ground parrot is thought to be one of the longest-living birds in the world. Also known as the owl parrot, it is nocturnal and can climb trees quite well, although it lives mostly in the underbrush. Found in New Zealand, the kākāpō is also important in Māori culture—the plentiful (at the time) birds were hunted for food, kept as pets, and valued for their beautiful feathers. One look at their wonderfully round, not-very-birdlike shape, endearing cheeks, and mustache *à la* Lorax and you can see why this charming ground parrot makes the ugly-cute cut.

FACT: The critically endangered parrots—whose numbers have been declining since European colonization—now live exclusively on four islands off the coast of New Zealand that have been cleared of predators. One kākāpō gained celebrity status after it attempted to mate with a zoologist's head during filming of the BBC program *Last Chance to See*, resulting in host Stephen Fry's delighted observation above. The bird was unsuccessful, but not for lack of trying. Sirocco's mating attempt went viral—the clip has been viewed about 22 million times. He is now the Official Spokesbird for Conservation in Australia and appears on TV and online to promote conservation efforts.

"It ought to be impossible to describe a creature as looking old-fashioned but that's exactly how [the Kākāpō] looks with his big sideburns and Victorian gentleman's face."

— **Stephen Fry**

Quiz:

Match the famous people to their ugly-cute pets

Quiz

Can you guess which famous person embraced which ugly-cute animal?

1 Salvador Dalí

2 Winston Churchill

3 Marie Antoinette

4 Dante Gabriel Rossetti

a Pugs

b Anteaters

c Wombats

d Platypus

Answers:

1b—Spanish artist Salvador Dalí loved anteaters. Many of the Surrealists did. Dalí was once famously photographed walking an anteater on a leash in Paris. He also had a pet ocelot, named Babou.

2d—Winston Churchill was smitten with the platypus, demanding that the Australian government send him one to keep as a pet. This was at the height of World War II, so you'd have hoped he had more important things to do. But he billed it as a morale-boosting thing—something the public would appreciate. The platypus died before it arrived, so it was a good thing the public never really found out about Churchill's plan until later. The resulting despair among the British people might have lost the war for the allies.

3a—Marie Antoinette was a pug lover. As a teenager, she asked for one to be sent from Vienna to her rooms in Versailles.

4c—The English author Rossetti had two pet wombats and wrote many poems about them.

Incidentally, Audrey Hepburn had a pet fawn. Make of that what you will. Some people just have no imagination.

Mary River Turtle

The Mary River turtle is one of the most punk-rock creatures in the natural world. With a green algae mohawk and two fleshy barbels that look like a devilish goatee, it's safe to say this anti-establishment turtle will fight for its right to party. They also breathe through their butts via a gilled orifice called a cloaca. Add that to their punk-rock resume.

FACT: This short-necked turtle was only discovered to be a distinct species in 1994, although they've been hanging around in some form or another for at least 40 million years. Because these turtles are so chill, as turtles tend to be, they became popular pets for kids in the 1960s and 70s, when pet shops sold them as penny turtles. Their nests along the single Australian river where they make their homes were pillaged to the point of endangerment. Now a protected species, these solitary turtles are quite shy, which adds to their sweet little rock vibe. If you're lucky enough to spot one while exploring the Mary River, say hello and move along.

Hagfish

The hagfish. Where to start? It looks like an uncooked sausage. Its skin is pink and loose, and it emits a kind of marine slime as a defence mechanism. It has a skull but no spine—the only living animal where this is known to be the case—and it can tie itself in knots to incapacitate predators. Oh, and it also eats dead things off the ocean floor by burrowing its face into the decaying flesh. They're sometimes referred to as snot snakes.

FACT: Unchanged for around 300 million years, these prehistoric, eel-like fish definitely aren't the type to get their own cutesy cartoon character or stuffed animal line. But the hagfish does have a few redeeming qualities—primarily, its super-soft slime. The slime contains lots of tiny protein threads and it turns out these soft, stretchy strands are a miracle of science. One expert described the slime as similar in complexity to a spiderweb, just found underwater. (And instantaneous—a small bit of slime can expand up to 10,000 times in under a second. Take that, spiders.) Out of the water, it's a thick, non-stick gel that can be manipulated and stretched. It's thought the slime may have medical uses as a hydrogel. The threads can also be used to create super-strong materials. In Korea, the edible substance is used as a substitute for egg whites, while cooked hagfish is a common seafood dish. If this still hasn't won you over to the wonders of the snot snake, just know that when they curl up they look sort-of cute. Like a little soft-pink spiral.

Dung Beetle

Dung beetles eat dung, as advertised. Depending on the species, they roll it up into little manure balls and take it elsewhere, or they simply bury it for safekeeping. Then they'll snack in peace on the nutrient-rich ... material. While no one would argue that dung-related things are in the least bit cute, this particular beetle does have a *dignified*, even charming, look. Like an all-black bumblebee, but with shiny armor. They're tiny tanks that can roll balls many times their own weight. And those balls? Perfectly spherical. Farmers love these little insects, in fact, because these hard-working gardeners improve soil quality, help with seed dispersal, and keep the fields tidy. But did you know they also follow the stars? The African dung beetle, *Scarabaeus satyrus*, uses the Milky Way to get its bearings before rolling its muck ball. *Which way shall I roll my dung? Aha! Follow that galaxy!* We know humans and birds use the stars to navigate, but now we know that insects do too.*

FACT: The scarab, an ancient form of the dung beetle, was sacred to the Egyptians and considered a symbol of rebirth and good fortune. Because the beetles ate dung as well as laid their eggs in it, they were thought to represent the circle of life. The dung ball represented the sun. Their image appears everywhere in ancient Egypt—there's a scarab hieroglyph, scarab statues, scarab jewelery and amulets, and scarab funerary art. Which just goes to show that even really crappy things can turn out to be truly amazing. So remember to be open to that possibility.

*Maybe they believe in astrology? That Daisy Dung Beetle, she's *such* a Gemini!

"Any foolish boy can stamp on a beetle, but all the professors in the world cannot make a beetle."

—**Arthur Schopenhauer**

California Condor

Nearly extinct by the late 1980s—numbering just 27 birds in total—North America's largest land bird is a conservation triumph. This may surprise cynics who assume that only the fuzziest, cutest animals get selected for saving, while the slimiest, weirdest ones get left to fend for themselves. The remaining birds were brought into captivity and bred before being released back into the wild over a period of decades, while new legislation aimed at protecting the birds from lead buckshot and chemicals like DDT helped the birds survive once released. Conservationists even trained the birds to avoid power lines. The next time someone complains about politicians not doing anything, you can point to the California Condor's existence as an example of good government at work.

FACT: These gigantic hissing and grunting vultures have featherless heads and necks and a wingspan of around 9.5 feet. They are carrion birds, so their diet consists of mostly dead stuff. Sort of gross, I guess. And yet their odd looks and cult fandom have translated into "condor cams," a popular phenomenon where people from across the world can watch these amazing birds going about their business. Thousands of condor-lovers, for example, tuned in to check in on baby Iniko, condor number 31, during a raging wildfire that threatened his nest. Luckily, Iniko survived, much to the relief of his many fans. Against all odds, the condor has been saved from extinction thanks to regular folks who took action. Now these strange-looking birds can be spotted in the Grand Canyon, Zion National Park, and across much of California. Beautiful things are possible when we work together.

Roti Island Snake-necked Turtle

These are turtles with extra-long necks. So imagine a classic turtle (cute), then imagine that turtle with a super-long neck and a head that looks like a snake (hmmmm). Please note that long neck is also bumpy and wart-covered. It's like millions of years ago it sort of half evolved and then got stuck, with one part of its body (neck and head) wanting to escape its shell and run free and the rest of it staying put. (Who can relate, right?) This turtle is found only in the wetlands of Rote Island, in Indonesia, and is one of the world's rarest species.

FACT: People want what they can't have, as the saying goes. The Roti Island snake-necked turtle was in such high demand back in the day that it was banned from the pet trade in 2001 in order to protect it from going extinct. Unfortunately, the turtle hasn't been seen in the wild since 2009, but there are some being bred in captivity. The world's most desired turtle, it turns out, is so amazing we've mostly destroyed it. Which is why it's always important to love things while we can. You never know when they may disappear.

Leaf Sheep Sea Slug

Sea slugs. Ocean caterpillars. The sheep of the sea. From above, they're nothing special, just another slow-moving bit of goo or plant matter. But look at one head on and you'll be smitten. These underwater slugs have tiny dots for eyes that cause them to resemble cartoon sheep. They're also topped with a little forest of leaf-like appendages that glow green in the water.

FACT: Found in the waters of Japan, Indonesia, and the Philippines, these are probably the cutest slugs known to humankind. Which is a fairly easy accomplishment, due to slugs mostly being, well, slugs. They're also the only non-plant organism in the entire world that can photosynthesize. They nosh a bit of algae and use the chloroplasts to turn sunlight into food, which can sustain them for months. They're the world's only solar-powered being. And while we can't eat it, a bit of sunlight is always good for us, too. Let's go take a walk, shall we?

Platypus

The platypus is the OG ugly-cute animal. The one everyone learned about in school. Found only in Australia, the first taxidermied specimen brought to Europe for study was assumed to be a fake—a bunch of animals sewn together. With a duck bill, a beaver tail, and feet like an otter, this egg-laying mammal confounded all who encountered it—all the Europeans, that is. For the Aboriginal people who'd been living with the platypus for thousands of years, it was just another perfectly unsurprising creature. Like the kangaroo or wombat. Another wombat likeness, the platypus is also found to glow in the dark. Is there anything this amazing animal can't do?

FACT: British Prime Minister Winston Churchill had a platypus obsession. During World War II, he for some reason requested, and was granted, the delivery of a single platypus from the Australian government—it was to be the only one of its kind to ever set foot in England. His Minister of Information was concerned, however, about logistics. "Now that you have achieved your ambition to possess a platypus, you must decide where you are going to house the little creature," the man wrote. "If you decide to keep it near you, you must send your cat Nelson into exile." Sadly, the sensitive platypus perished on the long journey and Nelson continued to rule the roost. Churchill certainly embraced one of the most important aspects of the BLOB Method: Be weird.

Fossa

Cryptoprocta ferox is the scientific name for the fossa. *Crypto* means hidden in ancient Greek, while *procta* means anus, and *ferox* means wild. Apparently these Madagascar carnivores have hard-to-find bum-holes, but surely there's more to them than just that? They like to hang out in trees! They're nocturnal! Sigh.

FACT: The fossa is like a baby leopard—if the leopard looked a bit like a mongoose. It's a disconcerting animal, to be sure. Go ahead and look at it fast—cat! Look again. Rodent, maybe? We know *cats* are cute, but this isn't that. It's unsettling when we can't put something neatly into its box, isn't it? Which is why the fossa is a classic ugly-cute animal. It's so close to being 100 percent cute, and yet it's not. What's an uptight person to do? For one thing, you can be glad you weren't named by a scientist. They seem to delight in giving these animals an inferiority complex.

Magnificent Frigatebird

These massive seabirds have a bright-red throat patch that inflates to an impressive size. Sort of like a bullfrog's throat, but on a giant scale. They spend most of their time flying high above the ocean, riding air currents, hardly ever coming down to Earth. They've been known to stay aloft for months at a time. In 1697, explorer William Dampier wrote, "It lives on Fish yet never lights on the water, but soars aloft like a Kite... His Wings are very long; his feet are like other Land-fowl, and he builds on Trees, where he finds any." So why doesn't this seabird alight on water, you might wonder? Turns out, it can't. It will drown.

FACT: These are one of the only seabirds in the world that can't get wet—they lack the waterproofing oil on their feathers that would keep them afloat. I guess we've all felt like life has dealt us a short hand sometimes. These seabirds have adapted pretty well to this evolutionary imperfection, though: they steal food from other birds to survive. That's life in the animal kingdom for you. Anyway, its big, scarlet throat pouch looks sort of like a giant red kidney bean—which in all fairness is pretty cute. Turns out even poorly behaved things have a few redeeming qualities. There's hope for some of us yet!

Quiz:

Match the ugly-cute animal to its natural habitat

Quiz

It's possible to learn things every once in a while, if we pay attention. With that in mind, here's a chance for you to up your cocktail party chatter by remembering a few facts about which ugly-cute animal came from where.

1 Fossa

2 California condor (look, we all deserve some easy answers once in a while. Life is hard enough without failing a silly quiz.)

3 Dung beetle

4 Wombat

5 Axolotl salamander

a California

b A Mexico City lake

c Madagascar

d Australia

e Egypt

Answers:

1 c
2 a
3 e
4 d
5 b

Rogues' gallery

(literally, more
gratuitous images of
cuglies for you to love ...)

Welcome to the best portrait gallery you'll ever have the opportunity to behold. Yes, the Louvre is lovely. As is the National Portrait Gallery in both London and Washington, D.C. We hear the Sistine Chapel is fine too—if you enjoy long lines and crowds. But this is a different sort of gallery. One made up of the traditionally shunned, the downtrodden, the at-risk-of-extinction and "never-heard-of-them" types of creatures. These are the faces of the best, most ugly-cutest creatures in the animal kingdom. Look upon them. And rejoice.

GOLDEN SNUB-NOSED MONKEY

"I'll say it until I'm blue in the face—beauty is in the eye of the beholder."

"There is nothing ugly; I never saw an ugly thing in my life; for let the form of an object be what it may— light, shade, and perspective will always make it beautiful."

—**John Constable**

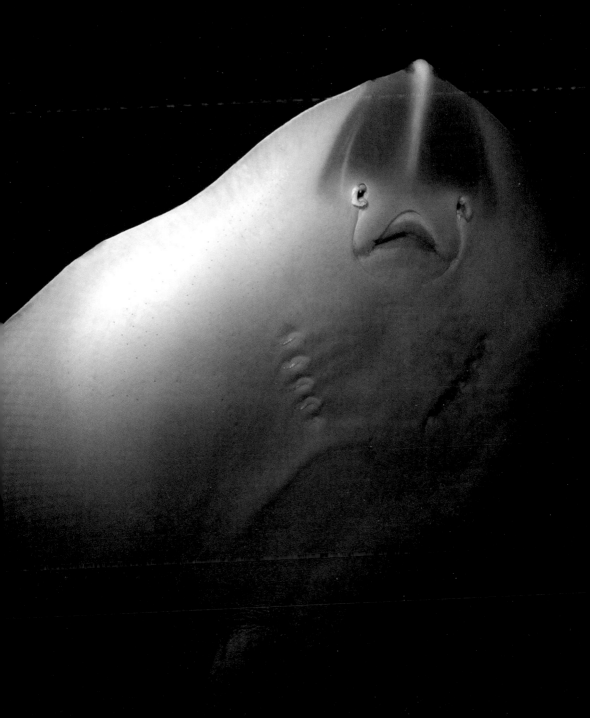

BROWN JUMPING SPIDER

Can a spider be cute? Well, it depends on who's asking! Just look at those fuzzy legs and all those wide eyes. This brown jumping spider is like a fluffy teddy bear—if that teddy bear was the size of your fingernail and had 8 legs ...

"Imperfection is beauty, madness is genius, and it's better to be absolutely ridiculous than absolutely boring."

–Marilyn Monroe*

*It's possible Monroe didn't say this. But like she may have said, imperfection is beauty, so we'll let it slide.

Philippine Tarsier

WILD TURKEY

An astonishing bird of even more astonishing proportions. These powerful birds are found in North and Central America—often roosting in trees which is slightly terrifying.

"It's quite simple: if someone says UR beautiful, believe them. If someone says UR ugly, don't believe them."

— **RuPaul**

SHOEBILL

This fellow has some evil Muppet energy.

Despite being docile creatures, their best defense is to simply glare.
Some truly passive-aggressive energy there.

"Beauty, to me, is about being comfortable in your own skin. That or a kick-ass red lipstick."

— Gwyneth Paltrow

Red-lipped Batfish

VENEZUELAN POODLE MOTH

A mystery moth. Photographed once in Venezuela and unlike any other moth science has seen before. Its fluffy appearance makes it look like a poodle or some other agreeable floof that might be nice to snuggle. Cute! But then again, imagine it in flight. This moth has some heft to it. Imagine it flapping about—this furry, bat-like, *Star Wars*-looking thing—bashing into your lamppost at night. And run for the hills.

"There is nothing in the world that is not mysterious, but the mystery is more evident in certain things than in others: in the sea, in the eyes of the elders, in the color yellow, and in music."

— Jorge Luis Borges

HONDURAN WHITE BAT

These fluffy popcorn bats have a triangle-shaped nose and are found in Central America.

"A big nose never spoiled
a handsome face."

— **French proverb**

Saiga Antelope

PYGMY HIPPO

Also known as "horse of the river"—although whether that nick-name was given by someone who had actually ever seen a horse is debatable. But adorably charming, whatever the moniker.

"Beauty is where you find it."

Baby White Rhinoceros

RACCOON

The raccoon (aka trash panda) is an urban icon. Look up at the trees or rooftops in places like Toronto or New York and you're likely to see one of these little bandits—probably eating some garbage.

"The best color in the whole world is the one that looks good on you."

— Coco Chanel

Shaggy Frogfish

UAKARI MONKEY

Like a sunburned toddler wearing a fur jacket. That is all.

"Everything has beauty, but
not everyone sees it."

— Confucius

South American Horned Frog, AKA Pacman Frog

WILDEBEEST

While wildebeest is a grand name, the alternative name "gnu" befits this charming mash-up of burly horns, skinny legs, and a delightful bulging belly.

"We live in a rainbow of chaos."

— Paul Cézanne

Panther Chameleon

COMMON POTOO

With a face like a recurring character on *Sesame Street*, the common potoo is also known as an urutau or poor-me-ones. Seriously. Related to nightjars and frogmouths, these nocturnal birds are notable for their massive eyes and huge, puppet-like mouth. If disturbed, the potoo "freezes" like a toddler playing hide-and-seek, by closing its eyelids, staring at the sky, and doing its best to keep still. Also, their eggs are white with lilac spots. What's not to love?

"Glow is the essence of beauty."

— Estée Lauder

Hawaiian Bobtail Squid

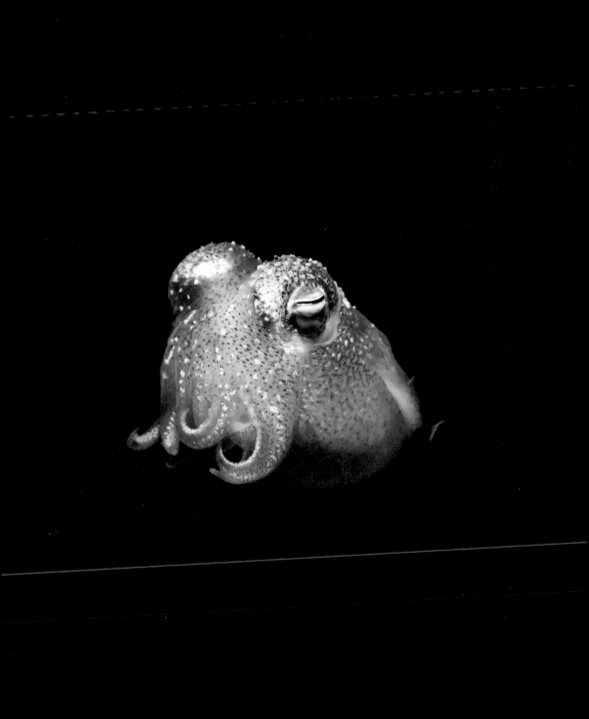

EMPEROR TAMARIN

Allegedly named for its resemblance to German emperor Wilhelm II for that mustachioed goodness.

If you ever climb to the top of a distant mountain, this is who is waiting for you at the summit. They will have some wise advice for you—well, most likely.

"The ugly can be beautiful.
The pretty, never."

— Oscar Wilde

Vietnamese Pot-bellied Pig

NAKED MOLE-RAT

The naked mole-rat is also called a sand puppy, which seems like some overly optimistic branding. If you're into petting animals that are tiny and hairless, though, this may be the pet for you.

"Take nothing for granted
as beautiful or ugly."

— **Frank Lloyd Wright**

Pig-nosed Turtle

LACKEY MOTH CATERPILLAR

Seriously, how is this even real?

While the lackey moth caterpillar is brightly colored with blue, orange, and white stripes, the adult moths are plain brown. But both the caterpillars and moths are sufficiently fuzzy-cute. The caterpillars like to hang out in groups. Picture feeding time at daycare, and you've got a lackey moth army. (Honestly, "army" is the official collective noun for caterpillars!)

"There's nothing wrong with being a large mammal."

—**Jim Morrison**

Walrus

ELEPHANT SHREW

Elephant shrews get their name from their longish trunk-like nose. Happily, their habitat is not (yet) threatened, and these little ones can be found thriving all over Africa.

"A full-grown manatee, which can weigh more than 1,000 pounds, looks like the result of a genetic experiment involving a walrus and the Goodyear Blimp."

— Dave Barry

Manatee

SOUTHERN FLANNEL MOTH CATERPILLAR

The nickname of woolly slug best suits this adorably fuzzy yet extremely venomous spiked caterpillar.

"If you are different from the rest of the flock, they bite you."

—Vincent O'Sullivan, *The Next Room*

Guinea Fowl

COMMON BUSHTAIL POSSUM

If you asked a five-year-old to draw a cat, squirrel, and fox at the same time, you'd get this charmer.

"It seems to me that what we call beauty in a face lies in a smile."

—Leo Tolstoy

Beluga Whale

TEXAS HORNED LIZARD

Cute face, horny spikes like a baby dragon—and the ability to shoot blood out of its eye sockets. The embodiment of ugly-cute.

"Confidence breeds beauty."

— Estée Lauder

Chinese Crested Dog

BABY AARDVARK

These little guys have the biggest, fleshiest ears. All the better to hear you with, my dear.

"Why should a lobster be any more ridiculous than a dog? Or a cat, or a gazelle, or a lion, or any other animal that one chooses to take for a walk? I have a liking for lobsters. They are peaceful, serious creatures."

— French writer Gérard de Nerval, ruminating on his pet lobster*

*It should be noted that when extolling the virtues of the lobster, de Nerval may not have been aware that it pees out of its face.

European Lobster

ORANGUTAN

You've heard of orangutans, obviously. Meaning "man of the jungle" in Malay, these shaggy, red-haired souls (adorned with something unfortunately called "face flanges") are highly intelligent. In fact, we share about 97% of our genes with these folks! How smart are they? One of the best orangutan ambassadors is Ken Allen. Ken is responsible for some of the greatest zoo escapes of all time. This San Diego-born Bornean orangutan escaped not once but three times from his enclosure. When zoo officials observed him in order to figure out how he did it, they surmised that the now-innocent-acting Ken knew he was being watched. Zookeepers ended up posing as tourists to try and catch him in the act but no dice. And what did Ken do when he escaped? He strolled around, visiting the other animals. Orangutans. They're just like us! Minus the face flanges, of course.

"What makes you different or weird, that's your strength."

—Meryl Streep

Pufferfish, AKA Blowfish

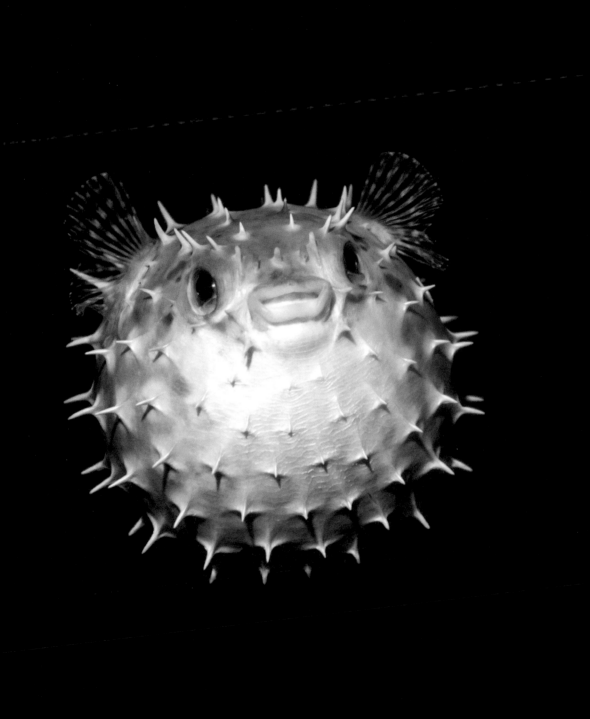

SILKIE CHICKEN

This is the sort of chicken you'd see at Fashion Week. Black skin, an "extra" chicken toe per foot, and the *pièce de résistance*? Blue earlobes.

"Here's what I think: the only reason I'm not ordinary is that no one else sees me that way."

—**R.J. Palacio,** *Wonder*

Yunnan Snub-nosed Monkey

CAPYBARA

Capybaras are, quite simply, guinea pigs who dreamed big. And then made that dream a reality.

"Who wants to be normal
when you can be unique?"

— Helena Bonham Carter

Lowland Streaked Tenrec

PORCUPINE

Porcupine means "quill pig" in Latin, although they're actually giant rodents. Their sweet faces and rotund, hair-covered bodies may look adorable, but the quills that cover their bodies are used for attack, and they eat bark, which can kill trees. So your view of porcupines depends firmly on whether you're a glass half full or glass half empty sort of person.

"The worse the haircut, the better
the man."

—**John Green**

Warthog

A very beautiful conclusion

> "Art produces ugly things which frequently become beautiful with time."
>
> **—Jean Cocteau**

Scientists reported in the confusingly titled journal *PLoS Biology* that there are an estimated 8.7 million different species in the natural world. We only *know* about a little over a million of them. The rest? A mystery. They have been living and growing and thriving and dying under the sea, in the trees, throughout our lakes, inside the tree bark, and everywhere all the time since the beginning of life itself. Without us ever even noticing or (gasp!) documenting them like the dutiful and curious scientists we are! Big and little. Weird and wonderful. Floofy and wet. All out there waiting to be discovered—or left alone! It's fine by them. They've done ok so far, right?

The animal world is magical and there's so much more to know. Gorillas weren't even described in print until 1847 and weren't studied until the 1920s. In 1951 birdwatchers spotted a bunch of "extinct" Bermuda petrels—the first sighting since the 1620s. The tiny leaf sheep slug was only discovered in 1993 by divers off the coast of Japan. Scientists found out in 2020 that platypus fur glows in the dark.

It is a miracle these creatures exist, and each one should be exalted. Or, at least, appreciated. Long tongues, skinny arms, rodent teeth— beautiful! Astonishing! Brilliant! Think what those appendages can accomplish. Finding food, scuffling home to see family, and gnawing on things that displease them. Using their perfect bodies and brains in the way nature and evolution intended. Animals— they're just like us. Ugly sometimes. Beautiful others. And all of us existing together on this fragile, resilient globe of space rock. If there was ever a better reason to keep an open mind and with- hold judgment, I haven't heard it.

So be open to the wonder. Be kind. Then next time you see a spider, try helping it on its way with a flick, rather than a smush. Everyone deserves a bit of compassion these days. And compassion looks cute on you.

A final thought

When life seems very boring or overwhelming, or when the world seems extra judgmental and unin- spired, remember the BLOB Method. Be weird. Love indiscriminately. Open up. And bury the hatchet. We could all learn to be a bit more like the blobfish.

Additional Resources

Conservation, education, and environmental organizations

The Ugly Animal Preservation Society
This society, founded in 2012, supports conservation efforts for the planet's less adorable creatures. Spot them giving educational talks, hosting comedy shows, and generally sticking up for the esthetically challenged all around the UK and abroad. Mascot? A blobfish. Excellent.
https://uglyanimalsoc.com/

The World Wildlife Fund (U.S. and Canada) or
The World Wide Fund for Nature (everywhere else)
This widely respected organization founded in 1961 supports wildlife conservation and endangered species across the globe. Its logo is, predictably, something very cute: a panda bear.
https://www.worldwildlife.org/

The National Geographic Society
Famous for its bright yellow magazines that are now in stacks in your garage, this society supports exploration, education, and conservation efforts around the world. Founded in 1888, it was one of the first publications in the world to print color photographs.

One of its reporting pledges is, rather charmingly: "Only what is of a kindly nature is printed about any country or people."
https://www.nationalgeographic.org/society/

Greenpeace

For those who feel that magazine reading isn't radical enough when it comes to saving the planet, there's Greenpeace. Traditionally this activist organization has focused on supporting actionable ways to save our environment and minimize the effects of the ongoing climate crisis. Whether it's reducing our use of single-use plastics, encouraging green initiatives in our cities, reducing our dependence on oil and gas, or getting to emissions zero—this is the place to go to make a difference.
https://www.greenpeace.org/international/

A very ugly-cute reading list

The Hunchback of Notre-Dame by Victor Hugo
This classic French historical novel from 1831 tells the story of the loving but misunderstood Quasimodo.

Frankenstein by Mary Shelley
This classic gothic novel is (loosely) about a misunderstood soul who just wants desperately to be loved. Blobfish much?

Slapstick by Kurt Vonnegut
A satire about two siblings shunned by their parents who come up with a plan to end loneliness in America. "If you can do no good, at least do no harm."

Anne of Green Gables by L. M. Montgomery
The classic novel about what it means to fit in, have a family, and be yourself.

Eleanor Oliphant Is Completely Fine by Gail Honeyman
Social outcast Eleanor forms an unlikely friendship with another lovable weirdo named Raymond. Everyone is worthy of love and friendship!

Index

About the Author

"I just do art because I'm ugly and
there's nothing else for me to do."
—Andy Warhol

Jennifer McCartney is a *New York Times* bestselling journalist,
editor, and author of numerous books. Her *Little Book of Animal
Philosophy* series is published in 17 countries. She has written
about utopias, time machines, and train travel for outlets like BBC
Radio 4, *The Atlantic*, *Architectural Digest*, *Vice Magazine*, *Teen
Vogue*, and CBC. She has also contributed to *Publishers Weekly*.
Originally from Hamilton, Ontario, she currently lives in Brooklyn,
New York with her two (cute) cats.

Acknowledgments

"You have this ability to find beauty in weird places."

—Kamila Shamsie

Many thanks to Caitlin Doyle at HarperCollins for her expert advice and patience with this lovely book. Thank you to my agent Euan Thorneycroft for all his hard work. And thanks to librarians and booksellers everywhere for being so wonderful.

Additionally, the publisher wishes to thank Jacqui Caulton for her excellent design work, as well as the editorial team who made this book possible. And to little Leonard, Mr Blobby's number-one fan.

Picture Credits

Photographs © Shutterstock.com with the exception of:

Pages 21 and 77 (blobfish) © Kerryn Parkinson / NORFANZ / Caters News / ZUMA Press

Pages 26 and 43 (sucker-footed bat) © Minden / naturepl.com

Pages 26 and 55 (yeti crab) © Fifis Alexis (2005). Galathea Yeti, an inhabitant of the abyssal depths. Ifremer. https://image.ifremer.fr/data/00569/68091/

Pages 26 and 56 (pug) © Claire Lloyd Davies

Pages 26 and 82 (kākāpō) © Minden Pictures / Alamy Stock Photo

Pages 26 and 91 (Mary River turtle) © Etienne Littlefair / naturepl.com

Page 28 (pink fairy armadillo) © Science History Images / Alamy Stock Photo

Page 36 (star-nosed mole) © Todd Pusser / naturepl.com

Page 70 (aquatic scrotum frog) © AIZAR RALDES/AFP via Getty Images

Page 85 (kākāpō) © Liu Yang / iStockphoto.com

Page 86 (Salvador Dalí) © Everett Collection / Bridgeman Images

Page 92 (hagfish) © Mark Conlin / Alamy Stock Photo

Page 166 (naked mole-rat) © Neil Bromhall / naturepl.com